Congressional Research Service

Navy CG(X) Cruiser Program: Background for Congress

Ronald O'Rourke
Specialist in Naval Affairs

June 10, 2010

Congressional Research Service

7-5700

www.crs.gov

RL34179

CRS Report for Congress
Prepared for Members and Committees of Congress

Summary

The Navy's FY2011 budget proposes canceling the CG(X) program as unaffordable and instead building an improved version of the Arleigh Burke (DDG-51) class Aegis destroyer called the Flight III version. This report provides background information on the CG(X) program as it existed prior to its proposed cancellation. For further discussion of the proposal to build Flight III DDG-51s in lieu of CG(X)s, see CRS Report RL32109, *Navy DDG-51 and DDG-1000 Destroyer Programs: Background and Issues for Congress*.

Contents

Appendixes

Contacts

Introduction

The Navy's FY2011 budget proposes canceling the CG(X) program as unaffordable and instead building an improved version of the Arleigh Burke (DDG-51) class Aegis destroyer called the Flight III version. This report provides background information on the CG(X) program as it existed prior to its proposed cancellation. For further discussion of the proposal to build Flight III DDG-51s in lieu of CG(X)s, see the CRS report on destroyer procurement.[1]

Background

CG(X) Cruiser Program Prior to Proposed Cancellation

This section briefly describes the CG(X) program as it existed prior to the proposal in the FY2011 budget to cancel the program and instead build Flight III DDG-51s.

Announcement of Program

The CG(X) program was announced on November 1, 2001, when the Navy stated that it was launching a Future Surface Combatant Program aimed at acquiring a family of next-generation surface combatants. This new family of surface combatants, the Navy stated, would include three new classes of ships:[2]

- **a destroyer called the DD(X)**—later renamed the DDG-1000 or Zumwalt class—for the precision long-range strike and naval gunfire mission,[3]

- **a cruiser called the CG(X)** for the AAW and BMD mission, and

- **a smaller combatant called the Littoral Combat Ship (LCS)** to counter submarines, small surface attack craft, and mines in heavily contested littoral (near-shore) areas.[4]

---

[1] CRS Report RL32109, *Navy DDG-51 and DDG-1000 Destroyer Programs: Background and Issues for Congress*, by Ronald O'Rourke.

[2] The Future Surface Combatant Program replaced an earlier Navy surface combatant acquisition effort, begun in the mid-1990s, called the Surface Combatant for the 21st Century (SC-21) program. The SC-21 program encompassed a planned destroyer called DD-21 and a planned cruiser called CG-21. When the Navy announced the Future Surface Combatant Program in 2001, development work on the DD-21 had been underway for several years, but the start of development work on the CG-21 was still years in the future. The DD(X) program, now called the DDG-1000 or Zumwalt-class program, is essentially a restructured continuation of the DD-21 program. The CG(X) might be considered the successor, in planning terms, of the CG-21. After November 1, 2001, the acronym SC-21 continued for a time to be used in the Navy's research and development account to designate a line item (i.e., program element) that funded development work on the DDG-1000 and CG(X).

[3] For more on the DDG-1000 program, see CRS Report RL32109, *Navy DDG-51 and DDG-1000 Destroyer Programs: Background and Issues for Congress*, by Ronald O'Rourke.

[4] For more on the LCS program, see CRS Report RL33741, *Navy Littoral Combat Ship (LCS) Program: Background, Issues, and Options for Congress*, by Ronald O'Rourke.

Replacement for CG-47s

The Navy wanted to procure as many as 19 CG(X)s as replacements for its 22 Ticonderoga (CG-47) class Aegis cruisers, which are projected to reach their retirement age of 35 years between 2021 and 2029.[5]

Planned Procurement Schedule

The Navy's FY2009 budget called for procuring the first CG(X) in FY2011. Beginning in late 2008, however, it was reported that the Navy had decided to defer the procurement of the first CG(X) by several years, to about FY2017.[6] Consistent with these press reports, on April 6, 2009, Secretary of Defense Robert Gates announced—as part of a series of decisions concerning the Department of Defense's (DOD's) proposed FY2010 defense budget—a decision to "delay the CG-X next generation cruiser program to revisit both the requirements and acquisition strategy" for the program.[7] The Navy's proposed FY2010 budget deferred procurement of the first CG(X) beyond FY2015.

Mission Orientation

The Navy's 22 Aegis cruisers are multi-mission ships with an emphasis on AAW and (for some CG-47s) BMD. The Navy similarly wanted the CG(X) to be a multi-mission ship with an emphasis on AAW and BMD. BMD has emerged in recent years as a significant new mission for the Navy.[8]

---

[5] CG-47s are equipped with the Aegis combat system and are therefore referred to as Aegis cruisers. A total of 27 CG-47s were procured for the Navy between FY1978 and FY1988; the ships entered service between 1983 and 1994. The first five, which were built to an earlier technical standard, were judged by the Navy to be too expensive to modernize and were removed from service in 2004-2005. The Navy is currently modernizing the remaining 22 to maintain their mission effectiveness to age 35; for more information, see CRS Report RS22595, *Navy Aegis Cruiser and Destroyer Modernization: Background and Issues for Congress*, by Ronald O'Rourke.

[6] Zachary M. Peterson, "Navy Awards Technology Company $128 Million Contract For CG(X) Work," *Inside the Navy*, October 27, 2008. Another press report (Katherine McIntire Peters, "Navy's Top Officer Sees Lessons in Shipbuilding Program Failures," *GovernmentExecutive.com*, September 24, 2008) quoted Admiral Gary Roughead, the Chief of Naval Operations, as saying: "What we will be able to do is take the technology from the DDG-1000, the capability and capacity that [will be achieved] as we build more DDG-51s, and [bring those] together around 2017 in a replacement ship for our cruisers." (Material in brackets in the press report.) Another press report (Zachary M. Peterson, "Part One of Overdue CG(X) AOA Sent to OSD, Second Part Coming Soon," *Inside the Navy*, September 29, 2008) quoted Vice Admiral Barry McCullough, the Deputy Chief of Naval Operations for Integration of Capabilities and Resources, as saying that the Navy did not budget for a CG(X) hull in its proposal for the Navy's budget under the FY2010-FY2015 Future Years Defense Plan (FYDP) to be submitted to Congress in early 2009.

An earlier report (Christopher P. Cavas, "DDG 1000 Destroyer Program Facing Major Cuts," *DefenseNews.com*, July 14, 2008) stated that the CG(X) would be delayed until FY2015 or later. See also Geoff Fein, "Navy Likely To Change CG(X)'s Procurement Schedule, Official Says," *Defense Daily*, June 24, 2008; Rebekah Gordon, "Navy Agrees CG(X) By FY-11 Won't Happen But Reveals Little Else," *Inside the Navy*, June 30, 2008.

[7] Source: Opening remarks of Secretary of Defense Robert Gates at an April 6, 2009, news conference on DOD decisions relating to DOD's proposed FY2010 defense budget.

[8] For further discussion, see CRS Report RL33745, *Navy Aegis Ballistic Missile Defense (BMD) Program: Background and Issues for Congress*, by Ronald O'Rourke.

Potential Design Features

The CG(X) was expected to feature a new radar, called the Air and Missile Defense Radar (AMDR), that would be larger and more powerful than the SPY-1 radar on the Navy's current Aegis cruisers and destroyers.[9]

The Navy originally intended to use its Zumwalt (DDG-1000) class destroyer hull design as the basis for the CG(X) design.[10] The potential for reusing the DDG-1000 hull design for the CG(X) was one of the Navy's arguments for moving ahead with the DDG-1000 program.[11] Subsequently, however, the Navy appeared to back away from the idea of reusing the DDG-1000 hull design as the basis for the CG(X).[12]

Section 1012 of the FY2008 defense authorization act (H.R. 4986/P.L. 110-181 of January 28, 2008) made it U.S. policy to construct the major combatant ships of the Navy, including ships like the CG(X), with integrated nuclear power systems, unless the Secretary of Defense submits a notification to Congress that the inclusion of an integrated nuclear power system is not in the national interest. The Navy studied nuclear power as a design option for the CG(X), but did not

---

[9] The Navy testified in 2007 that the power requirement of the CG(X) combat system, including the new radar, could be about 30 or 31 megawatts, compared with about 5 megawatts for the Aegis combat system. (Source: Spoken testimony of Navy officials to the Seapower and Expeditionary Forces Subcommittee of the House Armed Services Committee, March 1, 2007.) The CG(X) radar's greater power would be intended, among other things, to give the CG(X) more capability for BMD operations than Navy's Aegis cruisers and destroyers.

[10] For example, at an April 5, 2006, hearing, a Navy admiral in charge of shipbuilding programs, when asked what percentage of the CG(X) design would be common to that of the DDG-1000, stated that:

> [W]e haven't defined CG(X) in a way to give you a crisp answer to that question, because there are variations in weapons systems and sensors to go with that. But we're operating under the belief that the hull will fundamentally be—the hull mechanical and electrical piece of CG(X) will be the same, identical as DD(X). So the infrastructure that supports radar and communications gear into the integrated deckhouse would be the same fundamental structure and layout. I believe to accommodate the kinds of technologies CG(X) is thinking about arraying, you'd probably get 60 to 70 percent of the DD(X) hull and integrated (inaudible) common between DD(X) and CG(X), with the variation being in that last 35 percent for weapons and that sort of [thing]....

> The big difference [between CG(X) and DDG-1000] will likely [be] the size of the arrays for the radars; the numbers of communication apertures in the integrated deckhouse; a little bit of variation in the CIC [Combat Information Center—in other words, the] command and control center; [and] likely some variation in how many launchers of missiles you have versus the guns.

> (Source: Transcript of spoken testimony of Rear Admiral Charles Hamilton II, Program Executive Officer For Ships, Naval Sea Systems Command, before the Projection Forces Subcommittee of House Armed Services Committee, April 5, 2006. The inaudible comment may have been a reference to the DDG-1000's integrated electric-drive propulsion system. Between the two paragraphs quoted above, the questioner (Representative Gene Taylor) asked: "So the big difference [between CG(X) and DDG-1000] will be what?")

[11] For more on the DDG-1000, see CRS Report RL32109, *Navy DDG-51 and DDG-1000 Destroyer Programs: Background and Issues for Congress*, by Ronald O'Rourke.

[12] A July 2, 2008, letter from John Young, the Department of Defense (DOD) acquisition executive (the Under Secretary of Defense for Acquisition, Technology and Logistics) to Representative Gene Taylor, the chairman of the Seapower and Expeditionary Forces subcommittee of the House Armed Services Committee, stated: "I agree that the Navy's preliminary design analysis for the next-generation cruiser indicates that, for the most capable radar suites under consideration [for the CG(X)], the DDG-1000 [hull design] cannot support the radar." In addition, it is not clear that the DDG-1000 can accommodate one-half of the twin-reactor plant that the Navy has designed for its new Gerald R. Ford (CVN-78) class nuclear-powered aircraft carriers. If the DDG-1000 hull cannot accommodate one-half of the Ford-class plant, then the Navy might face a choice of either designing a new hull for the CG(X) that can accommodate one-half of the Ford-class plant or designing a new reactor plant that can fit into the DDG-1000 hull.

---

announce whether it would prefer to procure the CG(X) as a nuclear-powered ship. Some press reports suggested that a nuclear-powered version of the CG(X) might have a full load displacement of more than 20,000 tons and a unit procurement cost of $5 billion or more. The issue of nuclear power for Navy surface ships is discussed in more detail in another CRS report.[13]

Analysis of Alternatives (AOA)

The Navy assessed CG(X) design options in a study called the CG(X) Analysis of Alternatives (AOA), known more formally as the Maritime Air and Missile Defense of Joint Forces (MAMDJF) AOA. The CG(X) AOA was begun in mid-2006 and completed at the end of 2007. The Navy did not publicly release the results of the CG(X) AOA . **Appendix C** presents additional information on the CG(X) AOA.

FY2011 Proposal to Cancel CG(X) Program

The Navy's FY2011 budget proposes cancelling the CG(X) program and instead procuring an improved version of the DDG-51 called the Flight III version.[14] The Navy states that its desire to terminate the CG(X) program is "driven by affordability considerations."[15] Rather than starting to procure CG(X)s around FY2017, the Navy wants to begin procuring Flight III DDG-51s in FY2016. Navy plans thus call for procuring nine Flight IIA DDG-51s in FY2010-FY2015, followed by 24 Flight III DDG-51s between FY2016 and FY2031.[16]

The Flight III DDG-51 is to carry a version of the AMDR that is smaller and less powerful than the one envisaged for the CG(X). The Flight III DDG-51's AMDR is to have a diameter of about 14 feet, while the AMDR intended for the CG(X) might have had a diameter of about 22 feet.[17] In addition to improving the DDG-51's AAW and BMD capability through the installation of the AMDR, the Navy is also studying options for modifying the DDG-51 design in other ways for purposes of reducing crew size, achieving energy efficiency and improved power generation, improving effectiveness in warfare areas other than AAW and BMD, and reducing total

---

[13] CRS Report RL33946, *Navy Nuclear-Powered Surface Ships: Background, Issues, and Options for Congress*, by Ronald O'Rourke.

[14] It is a source of potential confusion that this is not the first time that the Navy has used the Flight III designation: The Navy in 1988 studied design options for a Flight III version of the DDG-51 design. The Chief of Naval Operations gave initial approval to a Flight III design concept, and the design was intended to begin procurement in FY1994. (Source: Donald Ewing, Randall Fortune, Brian Rochon, and Robert Scott, *DDG 51 Flight III Design Development*, Presented at the Meeting of the Chesapeake Section of The Society of Naval Architects and Marine Engineers, December 12, 1989.) The Flight III design was canceled in late-1990/early-1991. Subsequent studies led to the current Flight IIA design, which began procurement in FY1994. The Flight III DDG-51 that the Navy now wants to begin procuring in FY2016 is not the same as the Flight III design of 1988-1991.

[15] Department of the Navy, Office of Budget, *Highlights of the Department of the Navy FY 2011 Budget*, February 2010, p. 5-7.

[16] Source: Supplementary data on 30-year shipbuilding plan provided to CRS and the Congressional Budget Office (CBO) by the Navy on February 18, 2010.

[17] Sources for 14-foot and 22-foot figures: Zachary M. Peterson, "DDG-51 With Enhanced Radar in FY-16, Design Work To Begin Soon," *Inside the Navy*, February 8, 2010; Amy Butler, "STSS Prompts Shift in CG(X) Plans," *Aerospace Daily & Defense Report*, December 11, 2010: 1-2; "[Interview With] Vice Adm. Barry McCullough," *Defense News*, November 9, 2009: 38.

---

ownership cost.[18] Detailed design work on the Flight III DDG-51 will reportedly begin in FY2012 or FY2013.[19]

The Navy's desire to cancel the CG(X) and instead procure Flight III DDG-51s apparently took shape during 2009: at a June 16, 2009, hearing before the Seapower subcommittee of the Senate Armed Services Committee, the Navy testified that it was conducting a study on destroyer procurement options for FY2012 and beyond that was examining design options based on either the DDG-51 or DDG-1000 hull form.[20] A January 2009 memorandum from the Department of Defense acquisition executive had called for such a study.[21] In September and November 2009, it was reported that the Navy's study was examining how future requirements for AAW and BMD operations might be met by a DDG-51 or DDG-1000 hull equipped with a new radar.[22] On December 7, 2009, it was reported that the Navy wanted to cancel its planned CG(X) cruiser and instead procure an improved version of the DDG-51.[23] In addition to being concerned about the projected high cost and immature technologies of the CG(X),[24] the Navy reportedly had concluded that it does not need a surface combatant with a version of the AMDR as large and capable as the one envisaged for the CG(X) to adequately perform projected AAW and BMD missions, because the Navy will be able to augment data collected by surface combatant radars with data collected by space-based sensors. The Navy reportedly concluded that using data collected by other sensors would permit projected AAW and BMD missions to be performed adequately with a radar smaller enough to be fitted onto the DDG-51.[25] Reports suggested that the new smaller radar would be a scaled-down version of the AMDR originally intended for the CG(X).[26]

---

[18] Source: Memorandum dated February 2, 2010, from Director, Surface Warfare Division (N86) to Commander, Naval Sea Systems Command (SEA 05) on the subject "Technical Study In Support Of DDG 51 Class Resource Planning And Requirements Analysis," posted on InsideDefense.com (subscription required) February 19, 2009. See also Zachary M. Peterson, "Navy To Launch Technical Study And Cost Analysis For New DDG-51s," *Inside the Navy*, February 19, 2010.)

[19] Zachary M. Peterson, "DDG-51 With Enhanced Radar in FY-16, Design Work To Begin Soon," *Inside the Navy*, February 8, 2010.

[20] Source: Transcript of spoken remarks of Vice Admiral Bernard McCullough at a June 16, 2009, hearing on Navy force structure shipbuilding before the Seapower subcommittee of the Senate Armed Services Committee.

[21] A January 26, 2009, memorandum for the record from John Young, the then-DOD acquisition executive, stated that "The Navy proposed and OSD [the Office of the Secretary of Defense] agreed with modification to truncate the DDG-1000 Program to three ships in the FY 2010 budget submission." The memo proposed procuring one DDG-51 in FY2010 and two more FY2011, followed by the procurement in FY2012-FY2015 (in annual quantities of 1, 2, 1, 2) of a ship called the Future Surface Combatant (FSC) that could be based on either the DDG-51 design or the DDG-1000 design. The memorandum stated that the FSC might be equipped with a new type of radar, but the memorandum did not otherwise specify the FSC's capabilities. The memorandum stated that further analysis would support a decision on whether to base the FSC on the DDG-51 design or the DDG-1000 design. (Memorandum for the record dated January 26, 2009, from John Young, Under Secretary of Defense [Acquisition, Technology and Logistics], entitled "DDG 1000 Program Way Ahead," posted on InsideDefense.com [subscription required].)

[22] Zachary M. Peterson, "Navy Slated To Wrap Up Future Destroyer Hull And Radar Study," *Inside the Navy*, September 7, 2009. Christopher P. Cavas, "Next-Generation U.S. Warship Could Be Taking Shape," *Defense News*, November 2, 2009: 18, 20.

[23] Christopher J. Castelli, "Draft Shipbuilding Report Reveals Navy Is Killing CG(X) Cruiser Program," *Inside the Navy*, December 7, 2009.

[24] Christopher J. Castelli, "Draft Shipbuilding Report Reveals Navy Is Killing CG(X) Cruiser Program," *Inside the Navy*, December 7, 2009.

[25] Amy Butler, "STSS Prompts Shift in CG(X) Plans," *Aerospace Daily & Defense Report*, December 11, 2009: 1-2.

[26] Cid Standifer, "NAVSEA Plans To Solicit Contracts For Air And Missile Defense Radar," *Inside the Navy*, December 28, 2009; "Navy Issues RFP For Phase II of Air And Missile Defense Radar Effort," *Defense Daily*, (continued...)

The Navy's report on its FY2011 30-year (FY2011-FY2040) shipbuilding plan, submitted to Congress in conjunction with the FY2011 budget, states that the 30-year plan:

> Solidifies the DoN's [Department of the Navy's] long-term plans for Large Surface Combatants by truncating the DDG 1000 program, restarting the DDG 51 production line, and continuing the Advanced Missile Defense Radar (AMDR) development efforts. Over the past year, the Navy has conducted a study that concludes a DDG 51 hull form with an AMDR suite is the most cost-effective solution to fleet air and missile defense requirements over the near to mid-term....
>
> The Navy, in consultation with OSD, conducted a Radar/Hull Study for future destroyers. The objective of the study was to provide a recommendation for the total ship system solution required to provide Integrated Air and Missile Defense (IAMD) (simultaneous ballistic missile and anti-air warfare (AAW) defense) capability while balancing affordability with capacity. As a result of the study, the Navy is proceeding with the Air and Missile Defense Radar (AMDR) program....
>
> As discussed above, the DDG 51 production line has been restarted. While all of these new-start guided missile destroyers will be delivered with some BMD capability, those procured in FY 2016 and beyond will be purpose-built with BMD as a primary mission. While there is work to be done in determining its final design, it is envisioned that this DDG 51 class variant will have upgrades to radar and computing performance with the appropriate power generation capacity and cooling required by these enhancements. These upgraded DDG 51 class ships will be modifications of the current guided missile destroyer design that combine the best emerging technologies aimed at further increasing capabilities in the IAMD arena and providing a more effective bridge between today's capability and that originally planned for the CG(X). The ships reflected in this program have been priced based on continuation of the existing DDG 51 re-start program. Having recently completed the Hull and Radar Study, the Department is embarking on the requirements definition process for these AMDR destroyers and will adjust the pricing for these ships in future reports should that prove necessary.[27]

In testimony to the House and Senate Armed Services Committees on February 24 and 25, 2010, respectively, Admiral Gary Roughead, the Chief of Naval Operations, stated:

> Integrated Air and Missile Defense (IAMD) incorporates all aspects of air defense against ballistic, anti-ship, and overland cruise missiles. IAMD is vital to the protection of our force, and it is an integral part of our core capability to deter aggression through conventional means....
>
> To address the rapid proliferation of ballistic and anti-ship missiles and deep-water submarine threats, as well as increase the capacity of our multipurpose surface ships, we restarted production of our DDG 51 Arleigh Burke Class destroyers (Flight IIA series). These ships will be the first constructed with IAMD, providing much-needed Ballistic Missile Defense (BMD) capacity to the Fleet, and they will incorporate the hull, mechanical, and electrical alterations associated with our mature DDG modernization program. We will

---

(...continued)

December 24, 2009: 4.

[27] U.S. Navy, *Report to Congress on Annual Long-Range Plan for Construction of Naval Vessels for FY 2011*, February 2010, pp. 12, 13, 19. The first reprinted paragraph, taken from page 12, also occurs on page 3 as part of the executive summary.

---

spiral DDG 51 production to incorporate future integrated air and missile defense capabilities....

The Navy, in consultation with the Office of the Secretary of Defense, conducted a Radar/Hull Study for future surface combatants that analyzed the total ship system solution necessary to meet our IAMD requirements while balancing affordability and capacity in our surface Fleet. The study concluded that Navy should integrate the Air and Missile Defense Radar program S Band radar (AMDR-S), SPY-3 (X Band radar), and Aegis Advanced Capability Build (ACB) combat system into a DDG 51 hull. While our Radar/Hull Study indicated that both DDG 51 and DDG 1000 were able to support our preferred radar systems, leveraging the DDG 51 hull was the most affordable option. Accordingly, our FY 2011 budget cancels the next generation cruiser program due to projected high cost and risk in technology and design of this ship. I request your support as we invest in spiraling the capabilities of our DDG 51 Class from our Flight IIA Arleigh Burke ships to Flight III ships, which will be our future IAMD-capable surface combatant. We will procure the first Flight III ship in FY 2016.[28]

Legislative Activity for FY2011

FY2011 Defense Authorization Bill (H.R. 5136/S. 3454)

House

The House Armed Services Committee, in its report (H.Rept. 111-491 of May 21, 2010) on the FY2011 defense authorization bill (H.R. 5136), states that "the committee supports the Navy decision to re-start the DDG 51 class destroyer acquisition program and to work toward a flight III version of the vessel by fiscal year 2016."(Page 157) The report states at another point that:

> The committee is pleased with the effort by the Navy to undertake a comprehensive analysis of the radar and hull alternatives needed for a future sea-based ballistic missile defense (BMD) platform. The analysis has determined that the proposed Air and Missile Defense Radar (AMDR) system matched to a DDG 51 class destroyer hull is the most cost-effective method of fielding a new generation of sea-based BMD. The committee notes that this new radar development program will leverage existing technologies of both the DDG 1000 class destroyer program and the DDG 51 class destroyer program. The committee understands that the AMDR system is not likely to reach full development for a number of years and that a funding authorization request for the first ship will not occur until fiscal year 2016. (Pages 75-76)

Senate

The Senate Armed Services Committee, in its report (S.Rept. 111-201 of June 4, 2010) on the FY2011 defense authorization bill (S. 3454), states:

---

[28] Statement of Admiral Gary Roughead, Chief of Naval Operations, before the House Armed Services Committee on 24 February, 2010, pp. 10-11; and Statement of Admiral Gary Roughead, Chief of Naval Operations, before the Senate Armed Services Committee on 25 February, 2010, pp. 10-11.

The committee remains concerned with the Navy's ability to execute what it believes is an overly optimistic procurement strategy for large surface combatants. The truncation of the DDG–1000, the restart of the DDG–51 class and the proposed Flight III variant of the DDG–51 inject a great deal of instability into the SCN accounts. The Navy's testimony before Congress has led this committee to identify six risk areas in the Navy's plan for DDG–51s: (1) the availability of the Air and Missile Defense Radar; (2) the extent and cost of modifications to the underlying ship's design package to support proposed changes to the ship; (3) increased limitation on service life margins of the early restart ships; (4) combat system software integration; (5) the overall complexity of various separate programs that need to converge for successful completion of the restart and Flight III programs; and (6) cost and schedule growth for the Aegis Combat System Modernization. The committee expects the Navy to keep it closely apprised of developments in these risk areas so that it can monitor appropriate risk mitigation efforts. (Page 41)

For additional legislative activity concerning the Flight III DDG-51, see CRS Report RL32109, *Navy DDG-51 and DDG-1000 Destroyer Programs: Background and Issues for Congress.*

Appendix A. Legislative Activity in 2009

FY2010 Funding Request

The Navy's proposed FY2010 budget requested $340.0 million in research and development funding for the CG(X) program. Of this total, $190.0 million is for developing the CG(X)'s new radar (called the Air and Missile Defense Radar, or AMDR) and $150.0 million is for research and development work on the ship in general. The $190 million for the AMDR is Project 3186 (Air and Missile Defense Radar) of PE0604501N (Advanced Above Water Sensors). The $150 million for the CG(X) in general is PE0204201N (CG[X]).

FY2010 Defense Authorization Act (H.R. 2647/P.L. 111-84)

All of the legislative activity reported below on H.R. 2647/P.L. 111-84 occurred prior to the December 7, 2009, news report about the Navy's desire to cancel the CG(X) and instead procure improved DDG-51s.

House

The House Armed Services Committee, in its report (H.Rept. 111-166 of June 18, 2009) on H.R. 2647, recommends approving the Navy's FY2010 research and development funding requests for PE0604501N (Advanced Above Water Sensors) and PE0204201N (CG[X]) (page 168, line 105, and page 170, line 134). The report states:

> The committee supports the ongoing efforts to develop the next generation cruiser. The committee believes that the next generation cruiser must meet the challenge of emerging ballistic missile technology and that an integrated nuclear power system is required to achieve maximum capability of the vessel. (Page 72)

The report also states:

> The committee supports Navy research efforts to develop a radar system for the next generation cruiser (CGN(X)). The committee understands that ongoing analysis to determine radar sensitivity, power requirements, physical structure, and weight will dictate the size of the hull necessary for the vessel.
>
> Therefore the committee supports accelerated development of the combat system along with efforts to begin detailed design and construction of the vessel.
>
> The committee remains committed to the direction of section 1012 of the National Defense Authorization Act for Fiscal Year 2008 (Public Law 110–181), which requires the use of an integrated nuclear propulsion system for the CGN(X). (Page 75)

Senate

Division D of the FY2010 defense authorization bill (S. 1390) as reported by the Senate Armed Services Committee (S.Rept. 111-35 of July 2, 2009) presents the detailed line-item funding tables that in previous years have been included in the Senate Armed Services Committee's report on the defense authorization bill. Division D recommends increasing the Navy's funding request for PE0604501N (Advanced Above Water Sensors) by $50 million, with additional funding to be used for "mobile maritime sensor technology development" (page 677, line 105 of the printed bill), and recommends approving the Navy's funding request for PE0204201N (CG[X]) (page 678, line 134). The committee's report states:

> The budget request included $190.0 million in PE 64501N for development efforts in support of a next-generation cruiser, CG(X). CG(X) is planned to be the replacement for the CG–47 class cruiser, with primary missions including air and missile defense. The Navy's last long-range shipbuilding plan proposed to procure the first ship of the CG(X) program in 2011. That schedule was clearly too optimistic.
>
> Part of the delay came from questions about the CG(X) Analysis of Alternatives (AoA), called the Maritime Air and Missile Defense of Joint Forces (MAMDJF) AoA. One problem has been that demanding threat requirements have led to very demanding sensor requirements, some of which could only be fit on a cruiser-size vessel by achieving major technology breakthroughs.
>
> Another cause of the delay was that, as the committee understands it, the Secretary of the Navy was asking questions about potential contributions of off-board, networked sensors and why the MAMDJF vessel had to be self-sufficient for target acquisition and tracking.
>
> The committee recognizes that there are at least two other platforms within DOD inventories that could provide the basis for developing a more robust off-board sensor augmentation. Such an incremental development approach might not require that the Navy make such heroic technology improvements in surface combatant radar technology. These are the Navy's own programs to develop a Cobra Judy replacement vessel, and the Missile Defense Agency's Sea-Based X-Band radar.
>
> A mobile maritime sensor could improve upon the performance of either of these radars by making more modest technology improvements that could provide requisite capability for radars that would be less risky, cheaper to acquire and operate, and potentially available sooner than sensors that must provide equivalent performance from within the relatively constrained confines of a surface combatant.
>
> The committee recommends an increase of $50.0 million to: (1) develop a radar architecture that would provide full field of view; (2) design of a partial array prototype; (3) develop, build, and test components of such an array; and (4) fabricate and test a partial array prototype. Information resulting from such an effort could provide valuable information upon which to base informed decisions about the best way to support the maritime air and missile defense mission. (Pages 67-68)

Section 113 of S. 1390 would prohibit the obligation and expenditure of funds for the construction or advanced procurement of materials for surface combatants (including cruisers) procured after FY2011 until certain conditions are met, and would require DOD to submit certain reports. The text of Section 113 is as follows:

> SEC. 113. PROCUREMENT PROGRAMS FOR FUTURE NAVAL SURFACE COMBATANTS.
>
> (a) Limitation on Availability of Funds Pending Reports About Surface Combatant Shipbuilding Programs- The Secretary of the Navy may not obligate or expend funds for the construction of, or advanced procurement of materials for, a surface combatant to be

constructed after fiscal year 2011 until the Secretary has submitted to Congress each of the following:

(1) An acquisition strategy for such surface combatants that has been approved by the Department of Defense.

(2) The results of reviews by the Joint Requirements Oversight Council for an Acquisition Category I program that supports the need for an acquisition strategy to procure surface combatants after fiscal year 2011.

(3) A verification by an independent review panel convened by the Secretary of Defense that, in evaluating the shipbuilding program concerned, the Secretary of the Navy considered each of the following:

(A) Modeling and simulation, including war gaming conclusions regarding combat effectiveness for the selected ship platforms as compared to other reasonable alternative approaches.

(B) Assessments of platform operational availability.

(C) Life cycle costs from vessel manning levels to accomplish missions.

(4) An intelligence analysis reflecting a coordinated threat assessment of the Defense Intelligence Agency that provides the basis for deriving the mix of platforms in the shipbuilding program concerned when compared with the surface combatants in the 2009 shipbuilding plan.

(5) The differences in cost and schedule arising from the need to accommodate new sensors and weapons in future surface combatants to counter the future threats referred to in paragraph (4) when compared with the cost and schedule arising from the need to accommodate sensors and weapons on surface combatants as contemplated by the 2009 shipbuilding plan for the vessels concerned.

(6) A verification by the commanders of the combatant commands that the shipbuilding program for the vessels concerned would be preferable to the surface combatants included in the 2009 shipbuilding plan for the vessels concerned in meeting all of their future mission requirements.

(7) A joint review by the Navy and the Missile Defense Agency setting forth additional requirements for investment in Aegis ballistic missile defense (BMD) beyond the number of DDG-51 and CG-47 vessels planned to be equipped for this mission area in the budget of the President for fiscal year 2010 (as submitted to Congress pursuant to section 1105 of title 31, United States Code).

(b) Future Surface Combatant Acquisition Strategy- Not later than the date upon which President submits to Congress the budget for fiscal year 2012 (as so submitted), the Secretary of the Navy shall submit to the congressional defense committees a plan to provide for full and open competition on the combat systems for surface combatants proposed in the future-years defense program submitted to Congress under section 221 of title 10, United States Code, together with such budget. The plan shall include specifics on the intent of the Navy to satisfy criteria described in subsection (a) and evaluate applicable technologies during the request for proposal and selection process.

(c) Naval Surface Fire Support- Not later than 120 days after the enactment of this Act, the Secretary of the Navy shall submit to the congressional defense committees an update to the

March 2006 Report to Congress on Naval Surface Fire Support. The update shall identify how the Department of Defense intends to address any shortfalls between required naval surface fire support capability and the plan of the Navy to provide that capability. The update shall include addenda by the Chief of Naval Operations and Commandant of the Marine Corps, as was the case in the 2006 report.

(d) Technology Roadmap for Future Surface Combatants and Fleet Modernization-

(1) IN GENERAL- Not later than 120 days after the date of the enactment of this Act, the Secretary of the Navy shall develop a plan to incorporate into surface combatants constructed after 2011, and into fleet modernization programs, the technologies developed for the DDG-1000 destroyer and the DDG-51 and CG-47 Aegis ships, including the following:

(A) For the DDG-1000 destroyer—

(i) combat system;

(ii) multi-function and dual-band radars;

(iii) hull, mechanical and electrical systems achieving significant manpower savings; and

(iv) integrated electric propulsion technologies.

(B) For the DDG-51 and CG-47 Aegis ships—

(i) combat system, including missile defense capability;

(ii) hull, mechanical and electrical systems achieving manpower savings; and

(iii) anti-submarine warfare sensor systems designed for operating in open ocean areas.

(2) SCOPE OF PLAN- The plan required by paragraph (1) shall include sufficient detail for systems and subsystems to ensure that the plan—

(A) avoids redundant development for common functions;

(B) reflects implementation of Navy plans for achieving an open architecture for all naval surface combat systems; and

(C) fosters full and open competition.

(e) Definition- In this section:

(1) The term `2009 shipbuilding plan' means the 30-year shipbuilding plan submitted to Congress pursuant to section 231, title 10, United States Code, together with the budget of the President for fiscal year 2009 (as submitted to Congress pursuant to section 1105 of title 31, United States Code).

(2) The term `surface combatant' means a cruiser, a destroyer, or any naval vessel under a program currently designated as a future surface combatant program.

Regarding this section, the committee's report states:

The committee recommends a provision that would prevent the Navy from obligating any funds for building surface combatants after 2011 until the Navy conducts particular analyses, and completes certain tasks that should be required at the beginning of major defense acquisition programs (MDAP).

For at least the past couple of years, the Navy's strategy for modernizing the major surface combatants in the fleet has been in upheaval. The Navy was adamant that the next generation cruiser had to begin construction in the 2011-2012 timeframe. After 15 years of consistent, unequivocal support of the uniformed Navy for the fire support requirement, and for the DDG-1000 destroyer that was intended to meet that requirement (i.e., gun fire support for Marine Corps or Army forces ashore), the Navy leadership, in the middle of last year, decided that they should truncate the DDG-1000 destroyer program and buy DDG51 destroyers instead.

The Defense Department has announced that the Navy will complete construction of the three DDG–1000 vessels and will build three DDG–51 destroyers, one in fiscal year 2010 and two in fiscal year 2011. Beyond that, the plan is less well defined, and includes building only a notional ''future surface combatant,'' with requirements, capabilities, and costs to be determined.

Notwithstanding Navy protests to the contrary, this was mainly due to the Navy's affordability concerns. The committee notes with no little irony that this sudden change of heart on the DDG–1000 program is at odds with its own consistent testimony that ''stability'' in the shipbuilding programs is fundamental to controlling costs and protecting the industrial base.

The Navy claims the change of heart on the DDG–1000 program was related to an emerging need for additional missile defense capability that would be provided by DDG–51s and is being requested by the combatant commanders, and would be used to protect carrier battle groups against new threats.

The committee certainly believes that the services should have the ability to change course as the long-term situation dictates. However, since we are talking about the long-term and hundreds of billions of dollars of development and production costs for MDAPs, the committee believes that the Defense Department should exercise greater rigor in making sure such course corrections are made with full understanding of the alternatives and the implications of such decisions, rather than relying on inputs from a handful of individuals. The committee has only to look at the decision-making behind the major course correction in Navy shipbuilding that yielded the Littoral Combat Ship (LCS) to be concerned by that prospect.

Before deciding on a course of action regarding acquisition of surface combatants after 2011, we collectively have time to perform the due diligence that should be and must be performed at the beginning of any MDAP. That is what this section will ensure.

In addition, in order to deter any delaying action on conducting and completing the activities required by this section before 2011, the committee directs that the Secretary of the Navy obligate no more than 50 percent of the funds authorized for fiscal year 2010 in PE 24201N, CG(X), until the Navy submits a plan for implementing the requirements of this section to the congressional defense committees. (Pages 13-14; emphasis added)

Section 1012 of S. 1390 would repeal Section 1012 of the FY2008 defense authorization act (H.R. 4986/P.L. 110-181 of January 28, 2008). The committee's report states:

The committee recommends a provision [Section 1012] that would repeal section 1012 of the National Defense Authorization Act for Fiscal Year 2008 (P.L. 110-181).

Section 1012 of the National Defense Authorization Act for Fiscal Year 2008 (P.L. 110-181), as amended by section 1015 of the Duncan Hunter National Defense Authorization Act for Fiscal Year 2009 (P.L. 110-417), would require that all new classes of surface combatants and all new amphibious assault ships larger than 15,000 deadweight ton light ship displacement have integrated nuclear power systems, unless the Secretary of Defense determines that the inclusion of an integrated nuclear power system in such vessel is not in the national interest.

The committee believes that the Navy is already having too much difficulty in achieving the goal of a 313-ship fleet without adding a substantial increment to the acquisition price of a significant portion of the fleet. Moreover, current acquisition law and the Weapon System Acquisition Reform Act of 2009 (P.L. 111-23) emphasize the need to start acquisition programs on a sure footing as a central mechanism by which the Department of Defense (DOD) can get control of cost growth and schedule slippage on major defense acquisition programs. Therefore, Congress should be loathe to dictate a particular outcome of a requirements process before the Department has conducted the normal requirements review.

The committee expects that the Navy will continue to evaluate the integrated nuclear power alternative for any new class of major surface combatants, but would prefer that any Navy requirements analysis not be skewed toward a particular outcome. (Page 170)

Conference

The conference report (H.Rept. 111-288 of October 7, 2009) on H.R. 2647/P.L. 111-84 of October 28, 2009, authorizes an increase of $15 million to the Navy's funding request for PE0604501N (Advanced Above Water Sensors), with the additional funding to be used for "mobile maritime sensor technology development" (page 1004, line 105), and a decrease of $40 million to the Navy's funding request for PE0204201N (CG[X]), with the reduction being for "program delay." (Page 1006, line 134)

Section 125 prohibits the obligation and expenditure of funds for the construction or advanced procurement of materials for surface combatants (including cruisers) procured after FY2011 until certain conditions are met, and requires DOD to submit certain reports. The text of Section 125 is as follows:

> SEC. 125. PROCUREMENT PROGRAMS FOR FUTURE NAVAL SURFACE COMBATANTS.
>
> (a) LIMITATION ON AVAILABILITY OF FUNDS PENDING REPORTS ABOUT SURFACE COMBATANT SHIPBUILDING PROGRAMS.—The Secretary of the Navy may not obligate or expend funds for the construction of, or advanced procurement of materials for, a surface combatant to be constructed after fiscal year 2011 until the Secretary has submitted to Congress each of the following:
>
> (1) An acquisition strategy for such surface combatants that has been approved by the Under Secretary of Defense for Acquisition, Technology, and Logistics.
>
> (2) Certification that the Joint Requirements Oversight Council—
>
> (A) has been briefed on the acquisition strategy to procure such surface combatants; and

(B) has concurred that such strategy is the best preferred approach to deliver required capabilities to address future threats, as reflected in the latest assessment by the defense intelligence community.

(3) A verification by, and conclusions of, an independent review panel that, in evaluating the program or programs concerned, the Secretary of the Navy considered each of the following:

(A) Modeling and simulation, including war gaming conclusions regarding combat effectiveness for the selected ship platforms as compared to other reasonable alternative approaches.

(B) Assessments of platform operational availability.

(C) Life cycle costs, including vessel manning levels, to accomplish missions.

(D) The differences in cost and schedule arising from the need to accommodate new sensors and weapons in surface combatants to be constructed after fiscal year 2011 to counter the future threats referred to in paragraph (2), when compared with the cost and schedule arising from the need to accommodate sensors and weapons on surface combatants as contemplated by the 2009 shipbuilding plan for the vessels concerned.

(4) The conclusions of a joint review by the Secretary of the Navy and the Director of the Missile Defense Agency setting forth additional requirements for investment in Aegis ballistic missile defense beyond the number of DDG–51 and CG–47 vessels planned to be equipped for this mission area in the budget of the President for fiscal year 2010 (as submitted to Congress pursuant to section 1105 of title 31, United States Code).

(b) FUTURE SURFACE COMBATANT ACQUISITION STRATEGY.—Not later than the date upon which the President submits to Congress the budget for fiscal year 2012 (as so submitted), the Secretary of the Navy shall submit to the congressional defense committees an update to the open architecture report to Congress that reflects the Navy's combat systems acquisition plans for the surface combatants to be procured in fiscal year 2012 and fiscal years thereafter.

(c) NAVAL SURFACE FIRE SUPPORT.—Not later than 120 days after the enactment of this Act, the Secretary of the Navy shall submit to the congressional defense committees an update to the March 2006 Report to Congress on Naval Surface Fire Support. The update shall identify how the Department of Defense intends to address any shortfalls between required naval surface fire support capability and the plan of the Navy to provide that capability. The update shall include addenda by the Chief of Naval Operations and Commandant of the Marine Corps, as was the case in the 2006 report.

(d) TECHNOLOGY ROADMAP FOR FUTURE SURFACE COMBATANTS AND FLEET MODERNIZATION.—

(1) IN GENERAL.—Not later than 120 days after the date of the enactment of this Act, the Secretary of the Navy shall develop a plan to incorporate into surface combatants constructed after 2011, and into fleet modernization programs, the technologies developed for the DDG–1000 destroyer and the DDG–51 and CG–47 Aegis ships, including technologies and systems designed to achieve significant manpower savings.

(2) SCOPE OF PLAN.—The plan required by paragraph (1) shall include sufficient detail for systems and subsystems to ensure that the plan—

(A) avoids redundant development for common functions;

(B) reflects implementation of Navy plans for achieving an open architecture for all naval surface combat systems; and

(C) fosters competition.

(e) DEFINITIONS.—In this section:

(1) The term "2009 shipbuilding plan" means the 30-year shipbuilding plan submitted to Congress pursuant to section 231, title 10, United States Code, together with the budget of the President for fiscal year 2009 (as submitted to Congress pursuant to section 1105 of title 31, United States Code).

(2) The term "surface combatant" means a cruiser, a destroyer, or any naval vessel, excluding Littoral Combat Ships, under a program currently designated as a future surface combatant program.

Regarding Section 125, the conference report states that "the conferees agree to direct that the Secretary submit the plan for implementing the requirements of this section to the congressional defense committees at the same time as the President submits the budget request for fiscal year 2011." (Page 680)

Regarding Section 1012 of S. 1390 (see discussion above), the conference report states:

Repeal of policy relating to the major combatant vessels of the Unites States Navy

The Senate amendment contained a provision (sec. 1012) that would repeal section 1012 of the National Defense Authorization Act for Fiscal Year 2008 (Public Law 110–181). Section 1012, as amended, would require that all new classes of surface combatants and all new amphibious assault ships larger than 15,000 deadweight ton light ship displacement have integrated nuclear power systems, unless the Secretary of Defense determines that the inclusion of an integrated nuclear power system in such vessel is not in the national interest.

The House bill contained no similar provision.

The Senate recedes. (Page 822)

FY2010 DOD Appropriations Act (H.R. 3326/P.L. 111-118)

The House and Senate legislative activity reported below on H.R. 3326 occurred prior to the December 7, 2009, news report about the Navy's desire to cancel the CG(X) and instead procure improved DDG-51s.

House

The House Appropriations Committee, in its report (H.Rept. 111-230 of July 24, 2009) on H.R. 3326, recommends increasing the Navy's funding request for PE0604501N (Advanced Above Water Sensors) by $23 million, with the additional funding to be used for "Common Digital Sensor Architecture" ($3 million), "Submarine Navigation Decision Aids" ($5 million), and "Program Increase – Advanced Sensor Development" ($15 million) (page 257, line 105). The report recommends reducing the Navy's funding request for PE0204201N (CG[X]) by $40 million for "Program delay" (page 258, line 134).

Senate

The Senate Appropriations Committee, in its report (S.Rept. 111-74 of September 10, 2009) on H.R. 3326, recommends approving the Navy's funding request for PE0604501N (Advanced Above Water Sensors), and reducing the Navy's funding request for PE0204201N (CG[X]) by $64 million, of which $24 million is for "Propulsion development ahead of material solution decision" and $40 million is for "Unjustified request" (page 177, line 105 and page 184, line 134).

Final Version

In lieu of a conference report, the House Appropriations Committee on December 15, 2009, released an explanatory statement on a final version of H.R. 3326. This version was passed by the House on December 16, 2009, and by the Senate on December 19, 2009, and signed into law on December 19, 2009, as P.L. 111-118. The explanatory statement states on page 1 that it "is an explanation of the effects of Division A [of H.R. 3326], which makes appropriations for the Department of Defense for fiscal year 2010. As provided in Section 8124 of the consolidated bill, this explanatory statement shall have the same effect with respect to the allocation of funds and the implementation of this as if it were a joint explanatory statement of a committee of the conference."

The explanatory statement increases the Navy's funding request for PE0604501N (Advanced Above Water Sensors) by $16.4 million, with the additional funding to be used for "Common Digital Sensor Architecture" ($2.4 million), "Submarine Navigation Decision Aids" ($4 million), and "Program Increase – Advanced Sensor Development" ($10 million) (page 276, line 105). The explanatory statement reduces the Navy's funding request for PE0204201N (CG[X]) by $104 million, of which $40 million is for "Program delay," $24 million is for "Propulsion development ahead of material solution decision," and $40 million is for "Unjustified request" (page 278, line 134).

FY2009 Supplemental Appropriations Act (H.R. 2346/P.L. 111-32)

Senate

Section 308 of H.R. 2346 as passed by the Senate would rescind, among other things, $270.26 million in FY2009 funding for the Research, Development, Test and Evaluation, Navy (RDT&EN) appropriation account. This provision is also present in S. 1054 as reported by the Senate Appropriations Committee. The committee's report on S. 1054 (S.Rept. 111-20 of May 14, 2009, page 55) states that the $270.26 million includes a rescission of $100 million in FY2009 funding for the CG(X) program.

House

Section 10012 of H.R. 2346 as passed by the House would rescind, among other things, $30.51 million in FY2009 RDT&EN funding and $5 million in FY2008 RDT&EN funding, but the House Appropriation Committee's report on H.R. 2346 (H.Rept. 111-105 of May 12, 2009, page 32) states that these rescissions are for fuel and for a classified program, respectively, rather than for the CG(X) program.

Conference

Section 309 of the conference report (H.Rept. 111-151 of June 12, 2009) on H.R. 2346/P.L. 111-32 of June 24, 2009, includes a rescission of $73.6 million in FY2009 research and development funding for the CG(X) program. (Page 106)

Appendix B. FY2008 Defense Authorization Act Bill and Report Language

The FY2008 defense authorization bill was first reported by the House and Senate Armed Services Committees as H.R. 1585 and S. 1547, respectively. The president vetoed H.R. 1585 on December 28, 2007, citing to objections unrelated to the matters discussed in this CRS report. H.R. 1585 was succeeded by H.R. 4986, a bill that modified certain provisions of H.R. 1585 as to take into account the president's objections. H.R. 4986 was signed into law as P.L. 110-181 on January 28, 2008. For the parts of H.R. 4986 that are the same as H.R. 1585, including the matters discussed in this CRS report, the conference report on H.R. 1585 (H.Rept. 110-477 of December 6, 2008) in effect serves as the conference report for H.R. 4986.

House Report

The House Armed Services Committee, in its report (H.Rept. 110-146 of May 11, 2007) on H.R. 1585 stated the following:

> The committee believes that the mobility, endurance, and electric power generation capability of nuclear powered warships is essential to the next generation of Navy cruisers. The Navy's report to Congress on alternative propulsion methods for surface combatants and amphibious warfare ships, required by section 130 of the National Defense Authorization Act for Fiscal Year 2006 (P.L. 109-163), indicated that the total lifecycle cost for medium-sized nuclear surface combatants is equivalent to conventionally powered ships. The committee notes that this study only compared acquisition and maintenance costs and did not analyze the increased speed and endurance capability of nuclear powered vessels.
>
> The committee believes that the primary escort vessels for the Navy's fleet of aircraft carriers should have the same speed and endurance capability as the aircraft carrier. The committee also notes that surface combatants with nuclear propulsion systems would be more capable during independent operations because there would be no need for underway fuel replenishment. (Page 387)

Conference Report

Section 1012 of the conference report (H.Rept. 110-477 of December 6, 2007) on H.R. 1585 stated:

> SEC. 1012. POLICY RELATING TO MAJOR COMBATANT VESSELS OF THE STRIKE FORCES OF THE UNITED STATES NAVY.
>
> (a) INTEGRATED NUCLEAR POWER SYSTEMS.—It is the policy of the United States to construct the major combatant vessels of the strike forces of the United States Navy, including all new classes of such vessels, with integrated nuclear power systems.
>
> (b) REQUIREMENT TO REQUEST NUCLEAR VESSELS.—If a request is submitted to Congress in the budget for a fiscal year for construction of a new class of major combatant vessel for the strike forces of the United States, the request shall be for such a vessel with an integrated nuclear power system, unless the Secretary of Defense submits with the request a notification to Congress that the inclusion of an integrated nuclear power system in such vessel is not in the national interest.

(c) DEFINITIONS.—In this section:

(1) MAJOR COMBATANT VESSELS OF THE STRIKE FORCES OF THE UNITED STATES NAVY.—The term "major combatant vessels of the strike forces of the United States Navy" means the following:

(A) Submarines.

(B) Aircraft carriers.

(C) Cruisers, battleships, or other large surface combatants whose primary mission includes protection of carrier strike groups, expeditionary strike groups, and vessels comprising a sea base.

(2) INTEGRATED NUCLEAR POWER SYSTEM.—The term "integrated nuclear power system" means a ship engineering system that uses a naval nuclear reactor as its energy source and generates sufficient electric energy to provide power to the ship's electrical loads, including its combat systems and propulsion motors.

(3) BUDGET.—The term "budget" means the budget that is submitted to Congress by the President under section 1105(a) of title 31, United States Code.

Regarding Section 1012, the conference report stated:

The Navy's next opportunity to apply this guidance will be the next generation cruiser, or "CG(X)". Under the current future-years defense program (FYDP), the Navy plans to award the construction contract for CG(X) in fiscal year 2011. Under this provision, the next cruiser would be identified as "CGN(X)" to designate the ship as nuclear powered. Under the Navy's normal shipbuilding schedule for the two programs that already have nuclear power systems (aircraft carriers and submarines), the Navy seeks authorization and appropriations for long lead time nuclear components for ships 2 years prior to full authorization and appropriation for construction.

The conferees recognize that the milestone decision for the Navy's CG(X) is only months away. After that milestone decision, the Navy and its contractors will begin a significant design effort, and, in that process, will be making significant tradeoff decisions and discarding major options (such as propulsion alternatives). This is the normal process for the Navy and the Department of Defense (DOD) to make choices that will lead to producing a contract design that will be the basis for awarding the construction contract for the lead ship in 2011.

In order for the Navy to live by the spirit of this guidance, the conferees agree that:

(1) the Navy would be required to proceed through the contract design phase of the program with a comprehensive effort to design a CGN(X) independent of the outcome of decisions that the Navy regarding any preferred propulsion system for the next generation cruiser;

(2) if the Navy intends to maintain the schedule in the current FYDP and award a vessel in fiscal year 2011, the Navy would need to request advance procurement for nuclear components in the fiscal year 2009 budget request; and

(3) the Navy must consider options for:

(a) maintaining the segment of the industrial base that currently produces the conventionally powered destroyer and amphibious forces of the Navy;

(b) certifying yards which comprise that segment of the industrial base to build nuclear-powered vessels; or

(c) seeking other alternatives for building non-nuclear ships in the future if the Navy is only building nuclear-powered surface combatant ships for some period of time as it builds CGN(X) vessels; and

(d) identifying sources of funds to pay for the additional near-term costs of the integrated nuclear power system, either from offsets within the Navy's budget, from elsewhere within the Department's resources, or from gaining additional funds for DOD overall.

The conferees recognize that these considerations will require significant additional near-term investment by the Navy. Some in the Navy have asserted that, despite such added investment, the Navy would not be ready to award a shipbuilding contract for a CGN(X) in fiscal year 2011 as in the current FYDP.

Section 128 of the John Warner National Defense Authorization Act for Fiscal Year 2007 (P.L. 109-364) required that the Navy include nuclear power in its Analysis of Alternatives (AOA) for the CG(X) propulsion system. The conferees are aware that the CG(X) AOA is nearing completion, in which case the Navy should have some indications of what it will require to design and construct a CGN(X) class.

Accordingly, the conferees direct the Secretary of the Navy to submit a report to the congressional defense committees with the budget request for fiscal year 2009 providing the following information:

(1) the set of next generation cruiser characteristics, such as displacement and manning, which would be affected by the requirement for including an integrated nuclear power system;

(2) the Navy's estimate for additional costs to develop, design, and construct a CGN(X) to fill the requirement for the next generation cruiser, and the optimal phasing of those costs in order to deliver CGN(X) most affordably;

(3) the Navy's assessment of any effects on the delivery schedule for the first ship of the next generation cruiser class that would be associated with shifting the design to incorporate an integrated nuclear propulsion system, options for reducing or eliminating those schedule effects, and alternatives for meeting next generation cruiser requirements during any intervening period if the cruiser's full operational capability were delayed;

(4) the Navy's estimate for the cost associated with certifying those shipyards that currently produce conventionally powered surface combatants, to be capable of constructing and integrating a nuclear-powered combatant;

(5) any other potential effects on the Navy's 30-year shipbuilding plan as a result of implementing these factors;

(6) such other considerations that would need to be addressed in parallel with design and construction of a CGN(X) class, including any unique test and training facilities, facilities and infrastructure requirements for potential CGN(X) homeports, and environmental assessments that may require long-term coordination and planning; and

(7) an assessment of the highest risk areas associated with meeting this requirement, and the Navy's alternatives for mitigating such risk. (Pages 984-986)

Appendix C. CG(X) Analysis of Alternatives (AOA)

This appendix presents information about the CG(X) AOA

May 2009 Navy Testimony

The Navy testified on May 15, 2009, that:

> The Maritime Air and Missile Defense of Joint Forces (MAMDJF) Initial Capabilities Document (ICD) was validated by the Joint Requirements Oversight Council (JROC) in May 2006.
>
> The results of the Navy's Analysis of Alternatives (AoA) for the Maritime Air and Missile Defense of Joint Forces capability are currently within the Navy staffing process. Resulting requirements definition and acquisition plans, including schedule options and associated risks, are being evaluated in preparation for CG(X) Milestone A. This process includes recognition of the requirement of the FY 2008 National Defense Authorization Act, that all major combatant vessels of the United States Navy strike forces be constructed with an integrated nuclear power plant, unless the Secretary of Defense determines this not to be in the best interest of the United States.
>
> Vital research and development efforts are in progress for the Air and Missile Defense Radar which paces the ship platform development. Engineering development and integration efforts include systems engineering, analysis, computer program development, interface design, engineering development models, technical documentation, and system testing are in process to ensure a fully functional CG(X) system design.[29]

August 2009 GAO Letter Report

An August 2009 Government Accountability Office (GAO) letter report on the CG(X) AOA stated:

> In the CG(X) Analysis of Alternatives, the Navy identified six ship design concepts. These concepts include developing new designs as well as making modifications to previous hulls. For example, two concepts are based upon making modifications to the DDG 1000 Zumwalt-class destroyer and another concept is based upon making modifications to the DDG 51 Arleigh Burke-class destroyer. The ship design concepts vary in both capability, including the sensitivity of the radar and number of missile cells, and propulsion system. The variability is based on whether the concept uses a previous hull or is a new design. The Navy analyzed two new cruiser design concepts, one with a conventional propulsion system and one with a nuclear propulsion system. Both included the most sensitive radar and highest number of missile cells of all the concepts.
>
> The sensitivity of the radar on each ship design drives the ability of that ship to address threats that cause capability gaps for joint forces. The Navy developed a minimum

---

[29] Statement of the Honorable Sean J. Stackley, Assistant Secretary of the Navy, (Research, Development and Acquisition), and Vice Admiral Bernard J. McCullough, Deputy Chief of Naval Operations for Integration of Capabilities and Resources, Before the Subcommittee on Seapower and Expeditionary Forces of the House Armed Services Committee [Hearing] on Navy Force Structure and Shipbuilding, May 15, 2009, pp. 8-9.

performance standard that each alternative would need to meet to address the gap. As the radar sensitivity level increases, the capability gaps against these threats diminish because the radar's ability to meet the performance standards improves.[30]

---

[30] Government Accountability Office, *Defense Acquisitions: Additional Analysis Needed to Capture Cost Differences Between Conventional and Nuclear Propulsion for Navy's Future Cruiser*, GAO-09-886R, August 7, 2009, p. 1. GAO states that this letter report is an unclassified summary of a classified GAO report on the CG(X) AOA.

Appendix D. Earlier Oversight Issues for the CG(X)

This appendix presents potential oversight issues for Congress on the CG(X) program prior to the proposal in the FY2011 budget to cancel the CG(X) program and instead procure Flight III DDG-51s.

Prospects for Eight-Ship Program with One Ship Every Three Years

It was reported in February 2009 that the Navy was considering the option of reducing the CG(X) program to eight ships and procuring the ships at a rate of one ship every three years.[31] Assuming the first CG(X) is procured in FY2017, the eighth ship under such a profile would be procured in FY2038 and would enter service around 2044.

A potential oversight issue for Congress were the potential prospects for completing eight-ship program procured at a rate of one ship every three years. Skeptics might argue that there were at least three reasons why such a program with such a profile might not be pursued to completion:

- the 22-year period (FY2017-FY2038) over which the ships would be procured was a long-enough period of time that Navy spending priorities could change before all eight ships are procured;

- a procurement rate of one ship every three years could reduce production learning-curve benefits in the program, making the later ships in the program more expensive than they would be if the ships were procured more closely together; and

- a procurement rate of one ship every three years would mean that the last few ships in the program would enter service decades after the retirement of the Aegis cruisers that the ships are intended to replace, and potentially decades after the appearance of ASBMs and other threats that the ships are intended to counter.

Nuclear Power

A major issue for the CG(X) program was whether some or all CG(X)s should be nuclear-powered. As mentioned in the "Background" section, Section 1012 of the FY2008 defense authorization act (H.R. 4986/P.L. 110-181 of January 28, 2008) made it U.S. policy to construct the major combatant ships of the Navy, including the CG(X), with integrated nuclear power systems, unless the Secretary of Defense submits a notification to Congress that the inclusion of an integrated nuclear power system in a given class of ship is not in the national interest. The conference report on P.L. 110-181 contained extensive report language relating to Section 1012 (see **Appendix B**).

The Navy reported to Congress in January 2007 that equipping a notional ship broadly like the CG(X) with a nuclear power plant instead of a conventional (i.e., fossil-fuel) power plant would, other things held equal, increase the unit procurement cost of follow-on ships in the class by about $600 million to $700 million in constant FY2007 dollars. The report concluded that if oil

---

[31] Christopher P. Cavas, "U.S. May Cut 52 Ships From Plan," *Defense News*, February 16, 2009: 1, 12.

prices in coming years are high, much or all of the increase in unit procurement cost could be offset over the ship's service life by avoided fossil-fuel costs.

A nuclear-powered CG(X) would be more capable than a corresponding conventionally powered version because of the mobility advantages of nuclear propulsion, which include, for example, the ability to make long-distance transits at high speeds in response to distant contingencies without need for refueling. Navy officials also stated that a nuclear power plant might be appropriate for the CG(X) in light of the high energy requirements of the CG(X)'s powerful BMD-capable radar.[32]

The August 2009 GAO letter report on the CG(X) AOA stated:

> The draft cost analysis [in the CG(X) AOA]—which has not yet been approved within the Navy—includes a life-cycle cost estimate and a break-even analysis. The Navy estimated the life-cycle costs for 19 nuclear cruisers and 19 conventional cruisers using the 2007 price of crude oil. Then, in the break-even analysis, the Navy calculated the price of crude oil at which the cost of 19 nuclear cruisers equals the cost of 19 conventional cruisers. Using this analysis, the Navy determined that if oil prices behaved similarly to the past 35 years, the nuclear cruisers would be cheaper than the conventional cruisers. The Navy's analysis does not include: (1) present value analysis to adequately account for the decreasing time value of money, (2) alternative scenarios for the future price of oil, and (3) an examination of how a less efficient conventional propulsion system would affect its cost estimates. By incorporating present value analysis, as required by Department of Defense guidance, and future oil projections from the Department of Energy's Energy Information Administration, we found that the life-cycle cost of the conventional cruisers would be less than the nuclear cruisers. This demonstrates the sensitivity of the cost estimates to different assumptions, underscoring the need for more rigorous analysis before reaching conclusions about the alternatives.

> Recommendations for Executive Action

> We recommend that the Secretary of Defense require that the Navy (1) before finalizing Phase 2 of the Maritime Air and Missile Defense of Joint Forces Analysis of Alternatives, include present value analysis, alternative fuel scenarios, and analysis on the effect that a less efficient conventional propulsion system has on the cost estimates and (2) include present value analysis and alternative fuel scenarios in any future analyses of the trade-off between conventional and nuclear propulsion.

> Agency Comments

---

[32] See, for example, the comments of Rear Admiral Kevin McCoy at a June 25, 2007, conference in Arlington, VA, sponsored by the American Society of Naval Engineers (ASNE). A news article reporting McCoy's remarks stated in part:

> McCoy has cautioned that the [Navy's] alternate propulsion study [submitted to Congress in January 2007] is not a specific recommendation for using nuclear propulsion for the CG(X) cruisers, which are intended to perform missile defense.

> "Really the issue I'll tell you is not so much about the power plant but it's about the mission," McCoy said June 25. "And if you think the mission is sitting off a hostile coast looking for a BMD type mission for one-beam cycles on the big high-powered radar, we're talking the radar is costing in the 30 megawatts range. Then alternatives like nuclear power start to come in."

> (Emelie Rutherford, "Despite Hill Pressure, Navy Noncommittal On Nuclear Power For CG(X)," *Inside the Navy*, July 2, 2007.)

The Department of Defense provided us with restricted comments on our report. In its comments, the department agreed with the recommended actions. However, it disagreed with several of GAO's underlying analyses.[33]

For more on the issue of nuclear power for Navy surface ships, see CRS Report RL33946, *Navy Nuclear-Powered Surface Ships: Background, Issues, and Options for Congress*, by Ronald O'Rourke.

Technical Risk

The CG(X) was to use many new technologies being developed for the DDG-1000. A potential key technical risk specific to the CG(X) program concerned its powerful new BMD-capable radar. The need to reduce technical risk in the CG(X) radar may be one reason why the Navy deferred procurement of the lead CG(X) from FY2011 to FY2017. A November 29, 2007, press article reported that Rear Admiral Alan Hicks, the director of the Aegis ballistic missile defense (BMD) program, "cautioned" that:

> the Navy shouldn't attempt to go with a radically advanced radar for CG (X), at least not initially. Rather, he said, it might be wiser to go with incremental upgrades, steadily improving radar technology on the future cruiser that will take shape in the next decade, just as the existing Aegis system on cruisers and destroyers today has been upgraded steadily over two decades.
>
> "Lots of people want to build this incredible radar," Hicks said. On the one hand, he sees that as a valid eventual goal. But "I do believe you need to get there in a stepped function. Jumping to a radar that is three generations ahead in one leap is going to be terribly challenging, and may drive costs" skyward, imperiling the need to make CG (X) affordable, he said. "So we need to be very careful how we get a risk-reduction package to get to that cruiser," perhaps by using existing radar technology as a base to help reduce that development risk, he said, pointing to the success of the Aegis modernization program.[34]

Hull Design

In addition to the issue of nuclear power, another ship-design issue for the CG(X) was whether the ship should use the DDG-1000's tumblehome hull or some other hull. Potential alternative hulls included existing hulls such as the DDG-51 hull and the LPD-17 amphibious ship hull, both of which are conventional flared hulls, or a new flared hull design.

A tumblehome hull, with its reduced radar detectability, is viewed as useful for accomplishing the DDG-1000's mission of using its 155 mm guns to strike targets ashore—a mission that could require the DDG-1000 to operate fairly close to enemy shore-based radars. Some observers believed that a hull with reduced detectability is less critical for the CG(X), because the CG(X)'s AAW and BMD missions might not require it to approach enemy shores as closely, and because the energy radiating from the ship's powerful BMD-capable radar will in any event provide

---

[33] Government Accountability Office, *Defense Acquisitions: Additional Analysis Needed to Capture Cost Differences Between Conventional and Nuclear Propulsion for Navy's Future Cruiser*, GAO-09-886R, August 7, 2009, p. 2. GAO states that this letter report is an unclassified summary of a classified GAO report on the CG(X) AOA.

[34] Dave Ahearn, "Large Number of Aegis Ships Would Be Needed To Shield Europe: Admiral," *Defense Daily*, November 29, 2007.

enemy sensors with an indication of the ship's location. Other observers might argue that even if a ship's location is known, a hull with reduced detectability can improve the ship's ability to evade (or to use decoys to confuse) the homing devices in enemy anti-ship cruise missile and torpedoes, or the fusing mechanisms in enemy mines.

Even if the CG(X) did not require the reduced radar detectability of a tumblehome hull, reusing the DDG-1000's tumblehome hull for the CG(X) might still have had economic advantages in terms of avoiding the cost of designing a new hull (which could easily be in the hundreds of millions of dollars) and taking advantage of production learning-curve efficiencies achieved from earlier construction of DDG-1000s. Designing a new hull would have incurred hull-design costs and sacrifice the opportunity to take advantage of DDG-1000 production learning-curve benefits. On the other hand, a new-design hull might have more easily accommodated the power plant and combat system desired for the CG(X), and be designed with the latest features for reducing its production cost.

One option for making the CG(X) a nuclear-powered ship would have been to equip it with one-half of the new twin-reactor plant that the Navy has designed for its new Ford (CVN-78) class aircraft carriers.[35] Reusing the Ford-class reactor plant would have avoided the costs of developing a new reactor plant for the CG(X)—a cost that could exceed $1 billion.[36] The DDG-1000 hull (or an enlarged version of the DDG-51 hull) might have been too small to easily accommodate one-half of a Ford-class plant, at least not without making changes to the plant. Using one-half of the Ford-class plant without making changes to it might have required designing a new hull that is larger than the DDG-1000 hull. If so, then using one-half of the Ford-class plant would have posed a tradeoff between avoided reactor plant design costs and additional hull-design costs.

Unit Affordability vs. Unit Capability

Issues such as the question of nuclear power and the ship's hull design formed part of a more general potential general oversight issue for Congress concerning whether the Navy had achieved the best balance in the CG(X) design between unit affordability and unit capability. The CG(X) was one of the Navy's relatively few remaining opportunities to use a new ship design to manage the overall cost of the Navy's shipbuilding program. Navy officials were aware of this, but they also wanted the CG(X) to be capable of performing certain intended missions, including the BMD mission that drives the need for the CG(X) to carry a large and powerful new radar. Navy officials were seeking a design solution for the CG(X) that represented the best balance between unit affordability and unit capability. Achieving such a balance is a long-standing challenge in ship design.

BMD Impact on CG(X) Numbers and Schedule

An additional potential oversight issue for Congress concerned the possible effect of the BMD mission on the required number of CG(X)s and the schedule for procuring CG(X)s. The planned total of up to 19 CG(X)s reflected, in part, certain assumptions about the Navy's future role in

---

[35] For more on the Ford-class program, see CRS Report RS20643, *Navy Ford (CVN-78) Class Aircraft Carrier Program: Background and Issues for Congress*, by Ronald O'Rourke.

[36] The estimated development cost of the Ford-class plant is roughly $1.5 billion.

U.S. BMD operations. The Navy's future in U.S. BMD operations, however, had not yet been fully defined. It was possible that as the role became better defined, the total required number of CG(X)s could have changed.[37] A related question was whether the schedule for procuring CG(X)s was properly aligned with foreign-country ballistic missile development programs. A 2005 defense trade press report, for example, stated that "navy officials project" that China could field TBMs capable of hitting moving ships at sea by about 2015.[38]

Industrial-Base Implications

The question of whether some or all CG(X)s should be nuclear-powered had significant potential implications for the surface combatant industrial base because the two shipyards that have built all the Navy's cruisers and destroyers in recent years—GD/BIW and the Ingalls yard that forms part of NGSB—are not licensed to build nuclear-powered ships.[39]

The only two U.S. shipyards currently licensed to build nuclear-powered ships for the Navy are Newport News Shipbuilding of Newport News, VA, a part of NGSB, which builds nuclear-powered surface ships and submarines, and General Dynamics' Electric Boat Division (GD/EB) of Groton, CT, and Quonset Point, RI, which builds nuclear-powered submarines. These two yards have built every nuclear-powered ship procured for the Navy since FY1969.

There were at least three potential approaches for building nuclear-powered CG(X)s:

- Build them at Newport News, with GD/EB possibly contributing to the construction of the ships' nuclear portions.

- License GD/BIW and/or Ingalls to build nuclear-powered ships, and then build the CG(X)s at those yards.

- Build the nuclear portions of the CG(X)s at Newport News and/or GD/EB, the non-nuclear portions at GD/BIW and/or Ingalls, and perform final assembly, integration, and test work for the ships at either

    - Newport News and/or

      GD/EB, or

      GD/BIW and/or Ingalls.

---

[37] For more on this issue, see CRS Report RL33745, *Navy Aegis Ballistic Missile Defense (BMD) Program: Background and Issues for Congress*, by Ronald O'Rourke.

[38] Yihong Chang and Andrew Koch, "Is China Building A Carrier?" *Jane's Defence Weekly*, August 17, 2005. The article states that "navy officials project [that such missiles] could be capable of targeting US warships from sometime around 2015." A 2007 press report states that another observer believes that a MARV-equipped version of China's CSS-6 TBM may be close to initial operational status. (Bill Gertz, "Inside the Ring," *Washington Times*, July 20, 2007: 6. [Item entitled "New Chinese Missiles"]. The article stated that it was reporting information from forthcoming report on China's military from the International Assessment and Strategy Center authored by Richard Fisher.)

[39] GD/BIW has never built nuclear-powered ships, and has never been licensed to do so. The Ingalls yard within NGSS built nuclear-powered submarines until the early 1970s but is no longer licensed to build nuclear-powered ships. (Ingalls built 12 nuclear-powered submarines, the last being the Parche [SSN-683], which was procured in FY1968, entered service in 1974, and retired in 2005. Ingalls also overhauled or refueled 11 nuclear-powered submarines. Ingalls's nuclear facility was decommissioned in 1980.)

These options had significant potential implications for workloads and employment levels at each of these shipyards.

On the question of what would be needed to license Ingalls and/or GD/BIW to build nuclear-powered ships, the director of Naval Reactors (NR)—the office in charge of the Navy's nuclear propulsion program—testified in March 2007 that:

> Just the basics of what it takes to have a nuclear-certified yard, to build one from scratch, or even if one existed once upon a time as it did at Pascagoula, and we shut it down, first and foremost you have to have the facilities to do that. What that includes, and I have just some notes here, but such things as you have to have the docks and the dry-docks and the pier capability to support nuclear ships, whatever that would entail. You would have to have lifting and handling equipment, cranes, that type of thing; construction facilities to build the special nuclear components, and to store those components and protect them in the way that would be required.
>
> The construction facilities would be necessary for handling fuel and doing the fueling operations that would be necessary on the ship—those types of things. And then the second piece is, and probably the harder piece other than just kind of the brick-and-mortar type, is building the structures, the organizations in place to do that work, for instance, nuclear testing, specialized nuclear engineering, nuclear production work. If you look, for instance, at Northrop Grumman Newport News, right now, just to give you a perspective of the people you are talking about in those departments, it is on the order of 769 people in nuclear engineering; 308 people in the major lines of control department; 225 in nuclear quality assurance; and then almost 2,500 people who do nuclear production work. So all of those would have to be, you would have to find that workforce, certify and qualify them, to be able to do that.[40]

The director of NR testified that Newport News and GD/EB "have sufficient capacity to accommodate nuclear-powered surface ship construction, and therefore there is no need to make the substantial investment in time and dollars necessary to generate additional excess capacity."[41] In light of this, the Navy testified, only the first and third options above are "viable."[42] The director of NR testified that:

> my view of this is we have some additional capacity at both Electric Boat and at Northrop Grumman Newport News. My primary concern is if we are serious about building another nuclear-powered warship, a new class of warship, cost is obviously going to be some degree of concern, and certainly this additional costs, which would be—and I don't have a number to give you right now, but I think you can see it would be substantial to do it even if you could. It probably doesn't help our case to move down the path toward building another nuclear-powered case, when we have the capability existing already in those existing yards.[43]

---

[40] Spoken testimony of Admiral Kirkland Donald before the Seapower and Expeditionary Forces Subcommittee of the House Armed Services Committee, March 1, 2007.

[41] Statement of Admiral Kirkland H. Donald, U.S. Navy, Director, Naval Nuclear Propulsion Program, before the House Armed Services Committee Seapower and Expeditionary Forces Subcommittee on Nuclear Propulsion For Surface Ships, 1 March 2007, p. 13.

[42] Source: Statement of The Honorable Dr. Delores M. Etter, Assistant Secretary of the Navy (Research, Development and Acquisition), et al., before the Seapower and Expeditionary Forces Subcommittee of the House Armed Services Committee on Integrated Nuclear Power Systems for Future Naval Surface Combatants, March 1, 2007, p. 7.

[43] Spoken testimony of Admiral Kirkland Donald before the Seapower and Expeditionary Forces Subcommittee of the House Armed Services Committee, March 1, 2007.

With regard to the third option of building the nuclear portions of the ships at Newport News and/or GD/EB, and the non-nuclear portions at Ingalls and/or GD/BIW, the Navy testified that the "[l]ocation of final ship erection would require additional analysis." One Navy official, however, expressed a potential preference for performing final assembly, integration, and test work at Newport News or GD/EB, stating that:

> we are building warships in modular sections now. So if we were going to [ask], "Could you assemble this [ship], could you build modules of this ship in different yards and put it together in a nuclear-certified yard?", the answer is yes, definitely, and we do that today with the Virginia Class [submarine program]. As you know, we are barging modules of [that type of] submarine up and down the coast.
>
> What I would want is, and sort of following along with what [NR director] Admiral [Kirkland] Donald said, you would want the delivering yard to be the yard where the reactor plant was built, tooled, and tested, because they have the expertise to run through all of that nuclear work and test and certify the ship and take it out on sea trials.
>
> But the modules of the non-reactor plant, which is the rest of the ship, could be built theoretically at other yards and barged or transported in other fashion to the delivering shipyard. If I had to do it ideally, that is where I would probably start talking to my industry partners, because although we have six [large] shipyards [for building large navy ships], it is really two corporations [that own them], and those two corporations each own what is now a surface combatant shipyard and they each own a nuclear-capable shipyard. I would say if we were going to go do this, we would sit down with them and say, you know, from a corporation standpoint, what would be the best work flow? What would be the best place to construct modules? And how would you do the final assembly and testing of a nuclear-powered warship?[44]

For further discussion of the issue, see CRS Report RL33946, *Navy Nuclear-Powered Surface Ships: Background, Issues, and Options for Congress*, by Ronald O'Rourke.

Visibility of CG(X) Research and Development Costs

Another potential oversight issue for Congress was whether CG(X) research and development costs were sufficiently visible in Navy budget-justification documents. CG(X) research and development costs were found in the Research, Development, Test and Evaluation, Navy (RDTEN) appropriation account in:

- Program Element (PE) PE0204201N (CG[X]); and

- Project 3186 (Air and Missile Defense Radar) of PE0604501N (Advanced Above Water Sensors).

The entry for PE0204201N in the FY2010 budget-justification book for the RDTEN account stated that this PE is "a newly established PE for all CG (X) Research and Development" and that this PE "encompasses all CG (X) Projects." These statements could mislead readers into overlooking Project 3186 in PE0604501N, which accounted for the majority ($190 million) of the $340 million requested in FY2010 for work relating to the CG(X). The 11-page entry on

---

[44] Spoken testimony of Vice Admiral Paul E. Sullivan, Commander, Naval Sea Systems Command, to the Seapower and Expeditionary Forces Subcommittee of the House Armed Services Committee, March 1, 2007.

PE0204201N mentions Project 3186 on PE0604501N twice in tables that summarize "other program funding," but did not explain that this project funds the development of the AMDR.[45]

Author Contact Information

Ronald O'Rourke
Specialist in Naval Affairs
rorourke@crs.loc.gov, 7-7610

---

[45] The AMDR is intended not solely for the CG(X), but potentially for future destroyers as well. In this sense, Project 3186 is not strictly for the CG(X) program. Even so, Navy briefing materials on the Navy's proposed FY2010 budget include the $190 million for Project 3186 in the total amount requested for CG(X) research and development (see, for example, the briefing slide entitled "R&D Investment" in the Navy briefing entitled "Department of the Navy FY 2010 President's Budget, 18 May 2009, Rear Admiral J.T. Blake, Deputy Assistant Secretary of the Navy for Budget"), and May 2009 Navy testimony on Navy shipbuilding programs states, in the section on the CG(X) program, that "The FY 2010 President's Budget requests $190 million for the Air and Missile Defense Radar development and $150 million to continue maturation of the CG(X) design based on the preferred alternative selected." (Statement of the Honorable Sean J. Stackley, Assistant Secretary of the Navy, (Research, Development and Acquisition), and Vice Admiral Bernard J. McCullough, Deputy Chief of Naval Operations for Integration of Capabilities and Resources, Before the Subcommittee on Seapower and Expeditionary Forces of the House Armed Services Committee [Hearing] on Navy Force Structure and Shipbuilding, May 15, 2009, p. 9)

www.ingramcontent.com/pod-product-compliance
Lightning Source LLC
Chambersburg PA
CBHW081410170526
45166CB00010B/3281